MIDDLE GRADES MATHEMATICS PROJECT

Factors and Multiples

Activity Book

William Fitzgerald

Mary Jean Winter

Glenda Lappan

Elizabeth Phillips

Addison-Wesley Publishing Company

Menlo Park, California • Reading, Massachusetts • New York
Don Mills, Ontario • Wokingham, England • Amsterdam • Bonn
Paris • Milan • Madrid • Sydney • Singapore • Tokyo
Seoul • Taipei • Mexico City • San Juan

This book is published by Innovative Learning Publications.

ISBN 0-201-28760-9

2 3 4 5 6 7 8 9 10 - ML - 95 94 93 92 91

Contents

Introduction

Activity 1 **The Factor Game**
Materials 1-1, Factor Game **1**
Worksheet 1-1, 30-Game Board **2**
Worksheet 1-1, page 2 **3**
Worksheet 1-2, Analyzing First Moves for the 30-Game Board **4**
Worksheet 1-2, page 2 **5**
Worksheet 1-3, Moves for the 49-Game Board **6**
Worksheet 1-4, Practice Problems **7**
Worksheet 1-4, page 2 **8**

Activity 2 **Create a Game**
Materials 2-1, List of Possible Products Beginning at 1 **9**
Materials 2-2, 10 x 10 Product Game **10**
Worksheet 2-1, 6 x 6 Product Game **11**
Worksheet 2-2, Create a 3 x 3 Product Game **12**
Worksheet 2-3, Create a 4 x 4 Product Game **13**
Worksheet 2-4, Create a 5 x 5 Product Game **14**
Worksheet 2-5, Find the Factors **15**

Activity 3 **Factor Pairs**
Worksheet 3-1, Factor Pairs and Rectangles **16**
Worksheet 3-2, Crossing the Line **17**
Worksheet 3-3, Practice Problems **18**

Activity 4 **Factor Trees**
Worksheet 4-1, Product Puzzle I **19**
Worksheet 4-2, Factor Trees I **20**
Worksheet 4-3, Factor Trees II **21**
Worksheet 4-4, Multiplication Mazes **22**
Worksheet 4-5, Practice Problems **23**
Worksheet 4-6, Product Puzzle II **24**

Activity 5 **Common Multiples**
Worksheet 5-1, Least Common Multiples **25**
Worksheet 5-2, Practice Problems **26**
Worksheet 5-3, Applications of LCM **27**

Activity 6 **Common Factors**
Worksheet 6-1, Common Factors **28**
Worksheet 6-2, Greatest Common Factor **29**
Worksheet 6-3, Practice Problems **30**

Activity 7 **Sifting for Primes**

Worksheet 7-1, 100 Board **31**
Worksheet 7-2, 101–200 Board **32**
Worksheet 7-3, Using the Sieve **33**
Worksheet 7-4, Prime Puzzle **34**
Worksheet 7-5, Practice Problems **35**

Activity 8 **Paper Pool**

Materials 8-1, Basic Table **36**
Materials 8-2, Overlay 1 **37**
Materials 8-3, Overlay 2 **38**
Worksheet 8-1, Introduction to Paper Pool **39**
Worksheet 8-2, Paper Pool **40**
Worksheet 8-2, page 2 **41**
Worksheet 8-2, page 3 **42**
Worksheet 8-3, Record Sheet **43**
Worksheet 8-4, Supertables **44**
Worksheet 8-5, Advanced Paper Pool **45**
Worksheet 8-6, How Many Squares Are Crossed? **46**

Review Problems 47

To the Student

We hope you will share with us our excitement about mathematics. We are continually surprised and amazed by the interesting relationships and patterns that seem to be around us at every turn. For example, in the midst of studying the probability of success in shooting one-and-one foul shots in basketball, we suddenly find ourselves face-to-face with the concept of the Golden Ratio, which is closely related to the shape of the shell of a nautilus. The shooting success rate for a player to average one point in each one-and-one situation is the same as the change in the "bend" in the spiral of the shell of the mollusk as it grows—and for exactly the same mathematical reasons!

Many numerical ideas seem to have a geometric explanation, just as many geometrical notions can be described with arithmetic or algebra. Much of modern mathematical activity consists of trying to find metaphors or models of one idea or relationship in another apparently different setting. The early Greek mathematicians did not have algebra (as it wasn't invented yet), so they had to do all of their mathematics using geometry and drawing in sand tables. Classes were cancelled on windy days!

In today's world, mathematics is becoming necessary in wider and wider areas of life and work. As the world changes so quickly, new ways need to be developed to explain the changes. Computers allow us to ask questions that couldn't be asked a few years ago. Many tasks that used to be the main course of school mathematics are no longer necessary for humans to learn. We can spend our time and energy on much more creative aspects of learning and doing mathematics.

This book will allow you to investigate various mathematical relationships. Throughout, your teacher will be posing certain challenges that you will be exploring with other students in small groups. Solutions to these challenges may require you to conduct experiments, play and analyze mathematical games, construct models, or perform computer simulations that will then help you look for patterns, make conjectures, and test hypotheses. As you pursue these problems, if you continue to ask yourself what seems to be happening, look for patterns, form conjectures, and always ask yourself *why* your conjecture makes sense, you will be doing what mathematicians do.

As a result of these experiences, we hope you will come to understand and share with us the joy and wonder in the discovery of so many rich connections among these mathematical concepts, and their relationship to the world around us.

Factor Game

1	2	3	4	5
6	7	8	9	10
11	12	13	14	15
16	17	18	19	20
21	22	23	24	25
26	27	28	29	30

NAME ______________________

30–Game Board

1	2	3	4	5
6	7	8	9	10
11	12	13	14	15
16	17	18	19	20
21	22	23	24	25
26	27	28	29	30

30–Game Board

1	2	3	4	5
6	7	8	9	10
11	12	13	14	15
16	17	18	19	20
21	22	23	24	25
26	27	28	29	30

30–Game Board

1	2	3	4	5
6	7	8	9	10
11	12	13	14	15
16	17	18	19	20
21	22	23	24	25
26	27	28	29	30

NAME ____________________

30–Game Board

1	2	3	4	5
6	7	8	9	10
11	12	13	14	15
16	17	18	19	20
21	22	23	24	25
26	27	28	29	30

49–Game Board

1	2	3	4	5	6	7
8	9	10	11	12	13	14
15	16	17	18	19	20	21
22	23	24	25	26	27	28
29	30	31	32	33	34	35
36	37	38	39	40	41	42
43	44	45	46	47	48	49

NAME ____________________

Analyzing First Moves for the 30–Game Board

Factors | **First Moves**

1st number picked is	Opponent Gets	SUM	GOOD	NOT GOOD
2				
3				
4				
5				
6				
7				
8				
9				
10				
11				
12				
13				
14				
15				

NAME ______________________

Analyzing First Moves for the 30–Game Board

Factors | **First Moves**

1st number picked is	Opponent Gets	SUM	GOOD	NOT GOOD
16				
17				
18				
19				
20				
21				
22				
23				
24				
25				
26				
27				
28				
29				
30				

NAME

Moves for the 49–Game Board

Play the Factor Game on the 49–Game Board. Analyze the first moves by filling in the chart below. Since the numbers 1–30 have been done for the 30–Game Board you need only to look at the numbers 31–49.

Factors			First Moves	
1st number picked is	Opponent Gets	SUM	GOOD	NOT GOOD
31				
32				
33				
34				
35				
36				
37				
38				
39				
40				
41				
42				
43				
44				
45				
46				
47				
48				
49				

NAME ______________________

Practice Problems

Use the table of first moves for numbers 1 through 49 (Worksheet 1-3) to complete the following.

1. List all the prime numbers from 1 to 49.

2. List all the numbers from 1 to 49 that are abundant numbers.

3. List all the numbers from 1 to 49 that are deficient numbers.

4. List all the numbers from 1 to 49 that are perfect numbers.

5. List all the numbers that have 2 as a factor.

What do we call these numbers? ______________________

What do we call numbers that do not have 2 as a factor? ______________________

NAME ____________________

Practice Problems

6. What number(s) have the most factors?

7. What factor is paired with

16 to give 48? ____________________ 16 to give 32? ____________________

8. What factor is paired with

12 to give 48? ____________________ 12 to give 36? ____________________

9. What factor is paired with

6 to give 48? ____________________ 6 to give 42? ____________________

10. What factor is paired with

4 to give 48? ____________________ 4 to give 44? ____________________

11. What factor is paired with

2 to give 48? ____________________ 2 to give 34? ____________________

12. What is the *best* first move on a 49–Game Board?

13. What is the *worst* first move on a 49–Game Board?

List of Possible Products Beginning at 1

Factors Starting With 1	Possible Products	Number of Products Added	Total Number of Products
1	1	1	1
2	2, 4	2	3
3	3, 6, 9	3	6
4	8, 12, 16	3	9
5	5, 10, 15, 20, 25	5	14
6	18, 24, 30, 36	4	18
7	7, 14, 21, 28, 35, 42, 49	7	25
8	32, 40, 48, 56, 64	5	30
9	27, 45, 54, 63, 72, 81	6	36
10	50, 60, 70, 80, 90, 100	6	42
11	11, 22, 33, 44, 55, 66, 77, 88, 99, 110, 121	11	53
12	84, 96, 108, 120, 132, 144	6	59
13	13, 26, 39, 52, 65, 78, 91, 104, 117, 130, 143, 156, 169	13	72
14	98, 112, 126, 140, 154, 168, 182, 196	8	80
15	75, 105, 135, 150, 165, 180, 195, 210, 225	9	89
16	128, 160, 176, 192, 208, 224, 240, 256	8	97

This list is provided as an aid to the teacher. By examining the Total Number of Products column, reasonable board sizes can be determined. For example, a 7 × 7 board would produce 49 spaces. The table shows that there are 42 products with the factors 1 to 10, so seven spaces should be blacked out.

To read the chart, select a row and use all the factors and products from 1 through that row. For example, the row starting with 4 gives the factors 1, 2, 3, 4, and all their products: 1; 2, 4; 3, 6, 9; and 8, 12, 16.

10 × 10 Product Game

	1	2	3	4	5	6	7	8	
9	10	11	12	13	14	15	16	18	20
21	22	24	25	26	27	28	30	32	33
35	36	39	40	42	44	45	48	49	50
52	54	55	56	60	63	64	65	66	70
72	75	77	78	80	81	84	88	90	91
96	98	99	100	104	105	108	110	112	117
120	121	126	128	130	132	135	140	143	144
150	154	156	160	165	168	169	176	180	182
192	195	196	208	210	224	225	240	256	

Factors:

1 2 3 4 5 6 7 8

9 10 11 12 13 14 15 16

NAME ______________________

6 × 6 Product Game

1	2	3	4	5	6
7	8	9	10	12	14
15	16	18	20	21	24
25	27	28	30	32	35
36	40	42	45	48	49
54	56	63	64	72	81

Winner ______________________

1	2	3	4	5	6
7	8	9	10	12	14
15	16	18	20	21	24
25	27	28	30	32	35
36	40	42	45	48	49
54	56	63	64	72	81

Winner ______________________

1	2	3	4	5	6
7	8	9	10	12	14
15	16	18	20	21	24
25	27	28	30	32	35
36	40	42	45	48	49
54	56	63	64	72	81

Winner ______________________

1	2	3	4	5	6
7	8	9	10	12	14
15	16	18	20	21	24
25	27	28	30	32	35
36	40	42	45	48	49
54	56	63	64	72	81

Winner ______________________

1 2 3 4 5 6 7 8 9

NAME ____________________

Create a 3 × 3 Product Game

Factors	Possible Products

NAME ______________________

Create a 4 × 4 Product Game

Why are the corners blacked out?

NAME

Create a 5 × 5 Product Game

NAME ______________________

Find the Factors

Find the factors used to make these Product Game boards. List the factors in the spaces provided. Find the secret number (?).

1.

■	4	6
9	10	14
15	?	25
35	49	■

? = ☐

____ ____ ____ ____

2.

4	6	8	9
12	16	18	24
27	32	36	48
54	64	72	?

? = ☐

____ ____ ____ ____ ____ ____

NAME

Factor Pairs and Rectangles

1. 12 **Pairs**

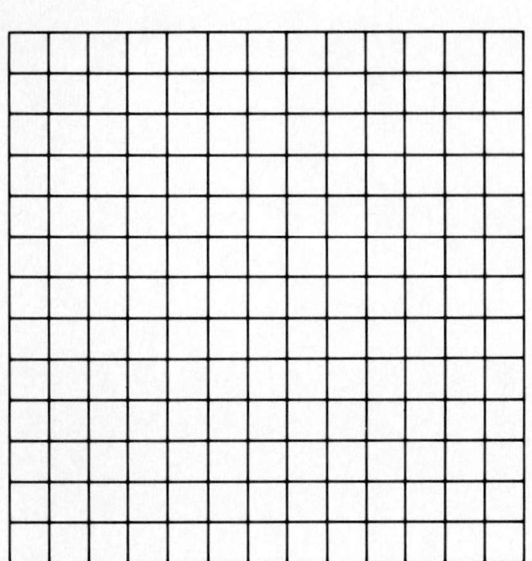

2. 20 **Pairs**

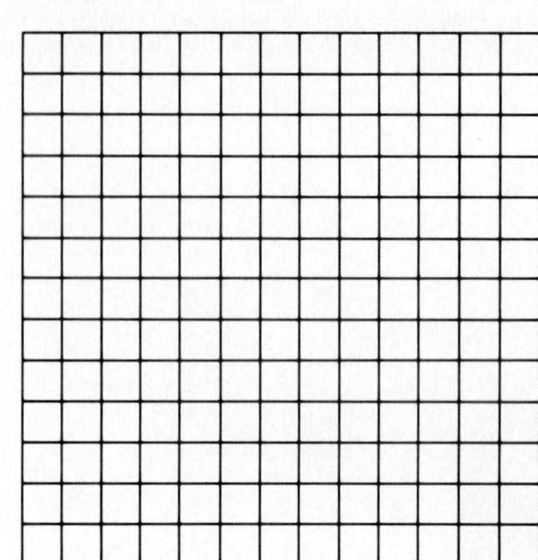

3. 36 **Pairs**

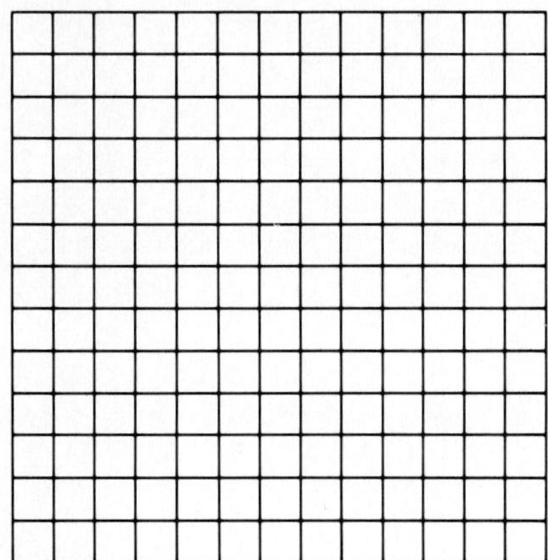

4. 30 **Pairs**

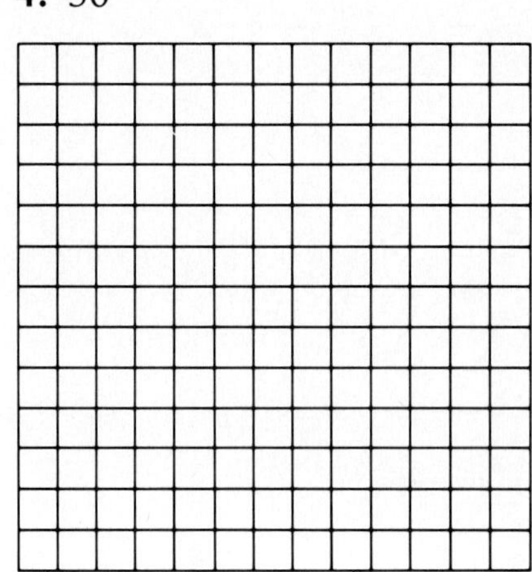

5. 40 **Pairs**

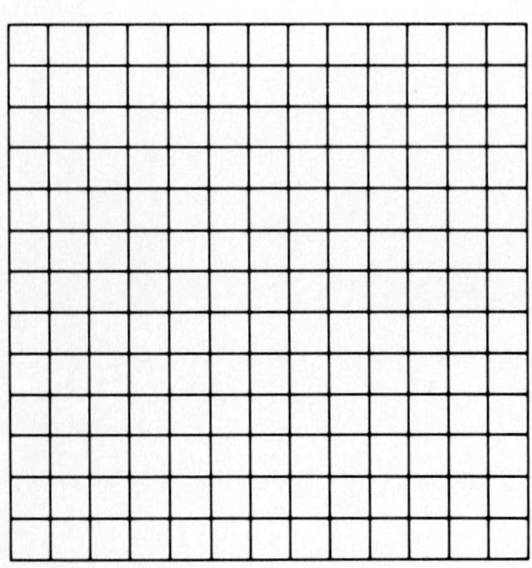

6. Squares **Draw the squares**

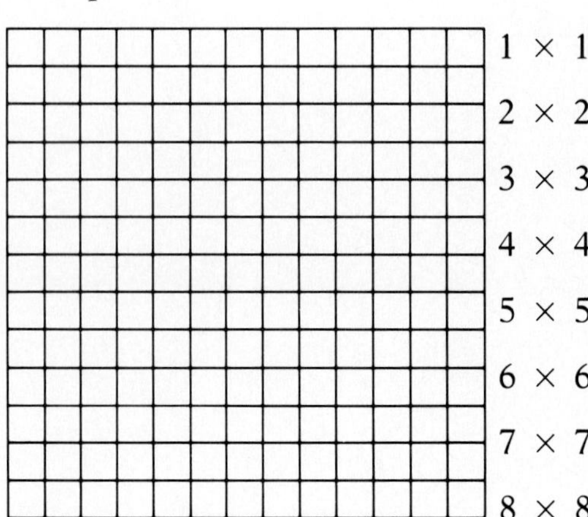

1 × 1
2 × 2
3 × 3
4 × 4
5 × 5
6 × 6
7 × 7
8 × 8

NAME ____________________

Crossing the Line

For each of the numbers, find where the factor pairs cross the line of squares.

1. 22

2. 44

3. 64

4. 128

5. 154

6. 278

7. To find all the factor pairs for 300, how far do you have to check to be sure that you have found them all?

8. List all the factor pairs for 300 (including the reversed ones).

NAME ______________________

Practice Problems

1. Mr. Brown wrote a number on the blackboard and said, "I know that to be sure I find all the factor pairs for this number, I must check each number from 1 to 47."

a) What is the smallest number Mr. Brown could have written on the blackboard?

b) What is the largest number Mr. Brown could have written on the blackboard?

2. Bob is making factor pairs of a number. He finds he can make exactly 7 factor pairs (including the reversed ones). If his number is less than 100, what is it?

3. Jo has chosen a number larger than 12 and less than 50. She finds exactly 6 factor pairs (including the reversed ones). What could her number be?

NAME ____________________

Product Puzzle I

Draw a loop around strings of numbers whose product is 1,350.

5	×	6	×	5	×	9	×	54
×	3	×	3	×	150	×	9	×
10	×	27	×	5	×	5	×	2
×	54	×	25	×	3	×	45	×
9	×	5	×	6	×	5	×	135
×	5	×	9	×	150	×	2	×
15	×	10	×	9	×	3	×	5
×	5	×	6	×	45	×	15	×
1	×	3	×	25	×	9	×	2

Record the different strings you have found.

1. ____________
2. ____________
3. ____________
4. ____________
5. ____________
6. ____________
7. ____________
8. ____________
9. ____________
10. ____________
11. ____________
12. ____________
13. ____________
14. ____________
15. ____________

After you have found 15 different strings, answer these questions.

1. Find 2 strings of numbers whose product is 1,350 that are not in the table.

2. Can you find a string whose product is 1,350 that is longer than any in the table?

NAME ______________________

Factor Trees I

For each of these numbers, grow a factor tree. When you reach the bottom of a tree, check it by multiplying the numbers together.

1. 36	**2.** 84
3. 72	**4.** 57
5. 144	**6.** 147
7. 63	**8.** 64

NAME ____________________

Factor Trees II

For each of these numbers, grow a factor tree. When you reach the bottom of a tree, check by multiplying the numbers together.

1. 840	**2.** 1,050
3. 109	**4.** 496

NAME ____________________

Multiplication Mazes

How can you use the fact that $840 = 2 \times 2 \times 2 \times 3 \times 5 \times 7$ to find the path through these mazes? No diagonals! Remember that the product of all the squares the path passes through must equal the exit number.

1. Find the path.

enter (top, above the 3)

5	3	10
6	7	2
11	2	4
	840 exit	

2. Find the *entrance* and the *path.*

14	2	2	
3	3	2	exit 840
5	7	6	

3. Make a factor tree for 3,927:

3,927

Find the entrance and the path.

8	5	11	
2	13	3	exit 3,927
11	17	7	

4. Make up your own numbers for this maze.

enter				
				5,775 exit

NAME

Practice Problems

1. Write the last four digits of your telephone number in order.

2. Find the prime factorization of the number you wrote in problem 1.

3. How many different prime factors are in your number?

4. How many prime factors greater than 13 are in your number?

5. Is your number prime?

6. Is your number divisible by 6?

7. Is your number divisible by 14?

8. Is your number divisible by 10?

9. Is your number divisible by 15?

NAME ____________________

Product Puzzle II

Find all strings whose product is 630. Strings may go horizontally, vertically, or bend around corners.

7	×	3	×	30
×	5	×	42	×
6	×	15	×	21
×	45	×	9	×
5	×	14	×	2

NAME ______________________

Least Common Multiples

In problems 1–3, use a calculator to find the multiples of each number. Write the least common multiple in the box. Show what each of the numbers must be multiplied by to get the LCM.

1. Multiples of 3: ______ ______ ______ ______ ______ ______ ______ ______

Multiples of 4: ______ ______ ______ ______ ______ ______ ______ ______

Common multiples of 3 and 4. ______________________________

3 × ______
4 × ______ = LCM = ☐

2. Multiples of 30: ______ ______ ______ ______ ______ ______ ______ ______

Multiples of 42: ______ ______ ______ ______ ______ ______ ______ ______

Common multiples of 30 and 42: ______________________________

30 × ______
42 × ______ = LCM = ☐

3. Multiples of 18: ______ ______ ______ ______ ______ ______ ______ ______

Multiples of 36: ______ ______ ______ ______ ______ ______ ______ ______

Common multiples of 18 and 36: ______________________________

18 × ______
36 × ______ = LCM = ☐

In problems 4–6, make a factor tree for each of the numbers. Use the prime factorizations to find the least common multiple.

4. 3 4 String for LCM ☐ = ______________

5. 30 40 String for LCM ☐ = ______________

6. 18 35 String for LCM ☐ = ______________

NAME ____________

Practice Problems

Find the least common multiple (LCM) of each pair of numbers given.

1. 4, 9 LCM is ______	**2.** 8, 14 LCM is ______	**3.** 10, 45 LCM is ______	**4.** 14, 15 LCM is ______
5. 14, 21 LCM is ______	**6.** 24, 36 LCM is ______	**7.** 58, 96 LCM is ______	**8.** 180, 210 LCM is ______

Find all the common multiples less than 100 for each pair of numbers given.

9. Common multiples of 5 and 7 ____________________

10. Common multiples of 12 and 20 ____________________

11. Common multiples of 4 and 6 ____________________

12. Common multiples of 9 and 15 ____________________

What is the smallest number that has the following factors?

13. 2 and 3.

14. 2, 3 and 4.

15. 4, 6 and 9.

16. 4, 6, 9 and 12.

17. I am thinking of a number. The least common multiple of my number and 9 is 45. What could my number be?

18. I am thinking of a number. My number has 8 as a factor and 12 as a factor.

a) What is the smallest that my number could be?

b) Name four other numbers that are factors of my number.

19. a) Find the LCM of 12 and 35.

b) Name 3 other common multiples of 12 and 35.

c) Name 3 other factors of the LCM.

NAME ______________________

Applications of LCM

1. Gleamy-Tooth toothpaste comes in 2 sizes.

GLEAMY-TOOTH — 9 oz for $0.89

GLEAMY-TOOTH — 12 oz for $1.19

a) What is the LCM of 9 and 12? ______________________

b) If you bought that much toothpaste in 9-oz tubes, how much would it cost? ______________________

c) If you bought that much toothpaste in 12-oz tubes, how much would it cost? ______________________

d) Which tube gives you more Gleamy-Tooth for the money?

____

2. In the school kitchen during lunch, the timer for pizza buzzes every 14 minutes; the timer for hamburger buns buzzes every 6 minutes. The two timers just buzzed together. In how many minutes will they buzz together again?

3. Two ships sail steadily between New York and London. One ship takes 12 days to make a round trip; the other takes 15 days. If they are both in New York today, in how many days will they both be in New York again?

4. The high school lunch menu repeats every 20 days; the elementary school menu repeats every 15 days. Both schools are serving sloppy joes today. In how many days will they both serve sloppy joes again?

5. Two neon signs are turned on at the same time. One blinks every 4 seconds; the other blinks every 6 seconds. How many times per minute do they blink on together?

6. How many teeth should be on gear A if each turn of gear A is to produce a whole number of turns of the shafts attached to B and C?

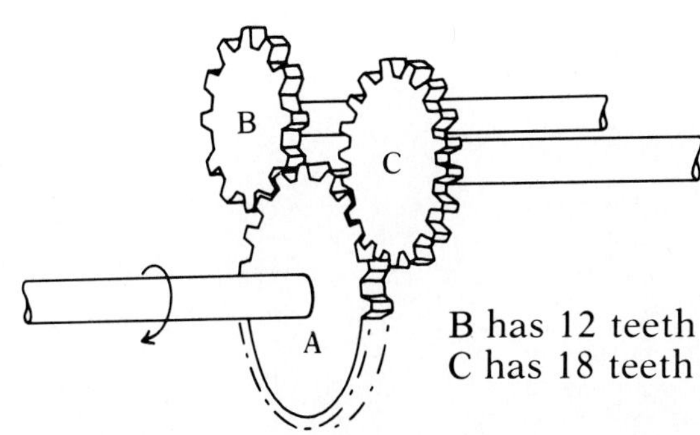

NAME ______________________

Common Factors

List the factors of each number. Then find the common factors of the two numbers. Do all your work in the space provided.

1. 9 and 24

2. 32 and 48

3. 51 and 17

4. 52 and 8

5. 1,001 and 70

6. 56 and 35

7. The scout leader has a certain number of cookies. They can be divided evenly among 9 scouts. They can also be divided evenly among 6 scouts. What are *two* possibilities for the number of cookies?

8. A band of pirates divided 185 pieces of silver and 148 gold coins. These pirates were known to be absolutely fair about sharing equally. How many pirates were there?

NAME ____________________

Greatest Common Factor

Solve problems 1–4 by finding the prime factorization of each number. Circle the largest string that is in both numbers. Then write the value of the greatest common factor in the box.

1. 56

36

GCF ☐

2. 18

24

GCF ☐

3. 36

19

GCF ☐

4. 52

13

GCF ☐

Use the following story to solve problems 5 and 6.

Ms. Wurst and Mr. Pop have donated a total of 91 hotdogs and 126 small cans of fruit juice for a math class picnic. Each student will receive the same amount of refreshments.

5. What is the greatest number of students that can attend the picnic?

How many cans of juice will each student receive?

How many hotdogs will each student receive?

6. If one of the hotdogs is eaten by Ms. Wurst's dog just before the picnic, what is the greatest number of students that can attend?

How many hotdogs will each student receive?

How many cans of juice will each student receive?

NAME ______________________

Practice Problems

List all the factors that the following sets of numbers have in common.

1. 21 and 49 ______________________

2. 17 and 37 ______________________

3. 12, 36, and 48 ______________________

4. 92 and 180 ______________________

What is the greatest common factor (GCF) of each of the following numbers?

5. 18 and 36 ______________________

6. 29 and 49 ______________________

7. 165 and 198 ______________________

8. 630 and 1,350 ______________________

9. What is the smallest number that 8 and 12 both divide?

10. a) If 8 and 20 both divide a number N, name four other numbers that must divide N. ______________________

    b) What is the smallest number that could be N? ______________________

NAME

100 Board

Code

1	2	3	4	5	6	7	8	9	10
11	12	13	14	15	16	17	18	19	20
21	22	23	24	25	26	27	28	29	30
31	32	33	34	35	36	37	38	39	40
41	42	43	44	45	46	47	48	49	50
51	52	53	54	55	56	57	58	59	60
61	62	63	64	65	66	67	68	69	70
71	72	73	74	75	76	77	78	79	80
81	82	83	84	85	86	87	88	89	90
91	92	93	94	95	96	97	98	99	100

NAME ____________________

101–200 Board

Code

101	102	103	104	105	106	107	108	109	110
111	112	113	114	115	116	117	118	119	120
121	122	123	124	125	126	127	128	129	130
131	132	133	134	135	136	137	138	139	140
141	142	143	144	145	146	147	148	149	150
151	152	153	154	155	156	157	158	159	160
161	162	163	164	165	166	167	168	169	170
171	172	173	174	175	176	177	178	179	180
181	182	183	184	185	186	187	188	189	190
191	192	193	194	195	196	197	198	199	200

NAME ______________________

Using the Sieve

Use your 100 Board to answer the following questions.

1. What is the smallest prime number that is greater than 30?

2. What is the smallest prime number that is greater than 50?

3. 5 and 7 are called *twin primes* because they are both primes and they differ by two. List all twin primes between 1 and 100.

4. A number that is not prime (and not 1) is a composite number. Find 5 composite numbers in a row.

5. Why didn't we sift for 9s?

6. Which of the primes 2, 3, 5, and 7 divide 84?

7. The number 6 was sifted with both 2 and 3.

a) Find all other numbers that were sifted with both 2 and 3.

b) How are all the multiples of 6 marked on the board?

8. a) List the multiples of 7.

b) What numbers are multiples of *both* 6 and 7.

9. How are multiples of 15 marked on the board?

10. There are four columns on the board that contain no primes. Find them and explain why these columns contain no primes.

NAME ____________________

Prime Puzzle

There is a message hidden below. Cross out the letters in the boxes containing numbers that are *not* prime numbers to discover the message in the remaining boxes.

D 7	P 6	I 2	R 8	V 19	I 11	M 12	P 60	S 3	K 9	S 14	O 59	Z 35	R 11	S 37
Q 4	A 3	R 31	M 25	E 23	S 10	D 29	M 12	I 41	V 97	H 100	I 23	N 83	E 13	A 12
B 71	U 2	R 35	T 3	T 27	F 43	O 42	A 37	I 64	C 7	T 5	R 45	O 13	R 11	S 71
N 9	E 14	U 69	M 32	A 17	S 87	F 48	G 75	O 20	R 19	K 9	E 97	Q 8	T 27	D 57
F 67	R 2	C 16	I 89	M 18	E 7	T 12	K 9	N 17	D 73	L 67	N 49	I 59	E 29	R 83

NAME ____________________

Practice Problems

Use your 100–200 Board to answer the following questions.

1. List all the primes between 100 and 200.

2. What prime numbers are factors of 84?

3. What prime numbers are factors of 110?

4. List all the multiples of 7 that are between 130 and 200.

5. Why didn't we sift for 17s on the 200 board?

6. A number that is not prime (and not 1) is a composite number. Find 5 composite numbers in a row on the 200 board. What is the longest string you can find?

7. Why is it impossible to find a number on the 200 board that is divisible by four different prime numbers? (Hint: How big would such a number have to be?)

8. What would be the last prime we would have to sift to find all primes less than 1,000?

Basic Table

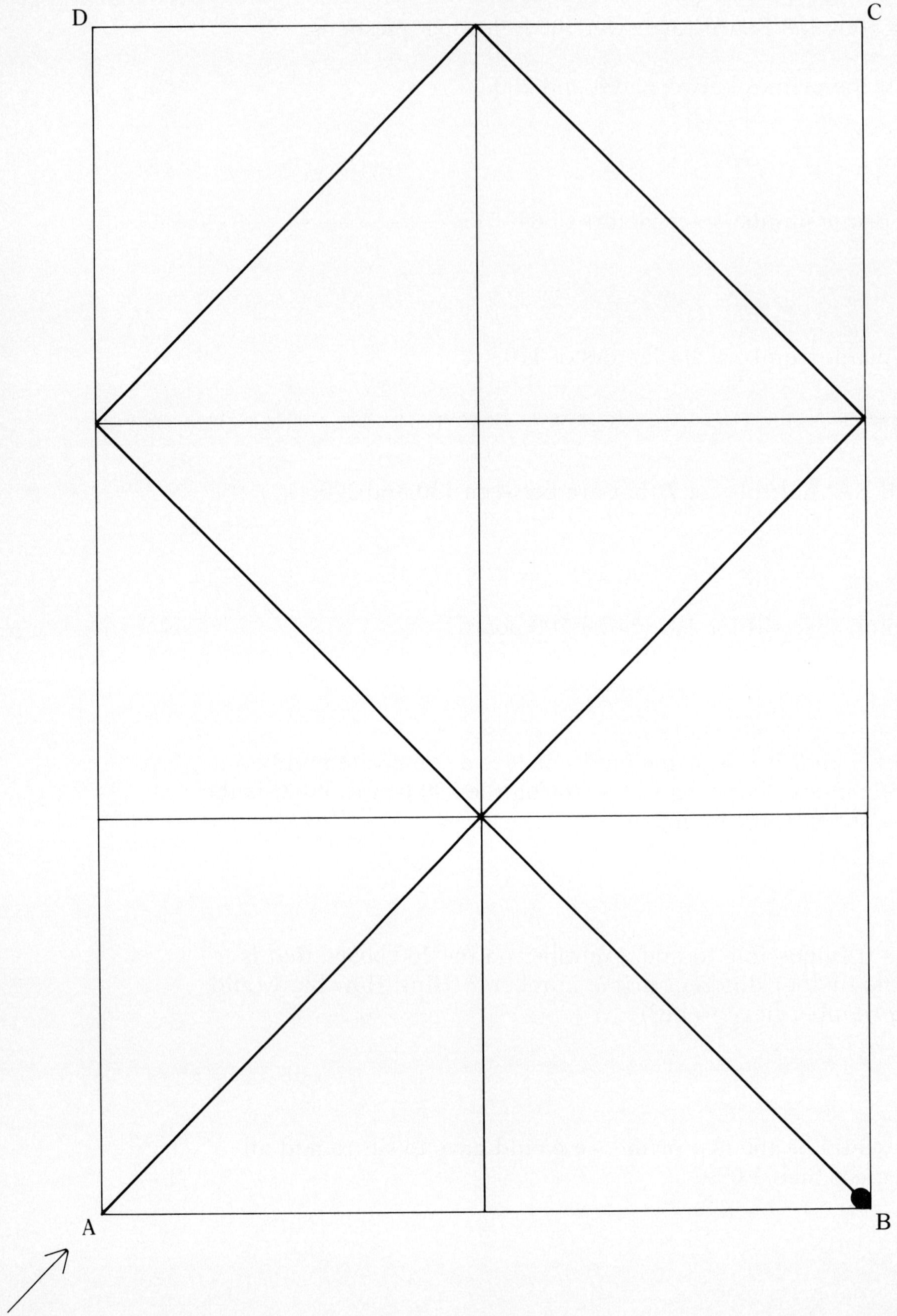

D

C

A

B

Overlay 1

Overlay 2

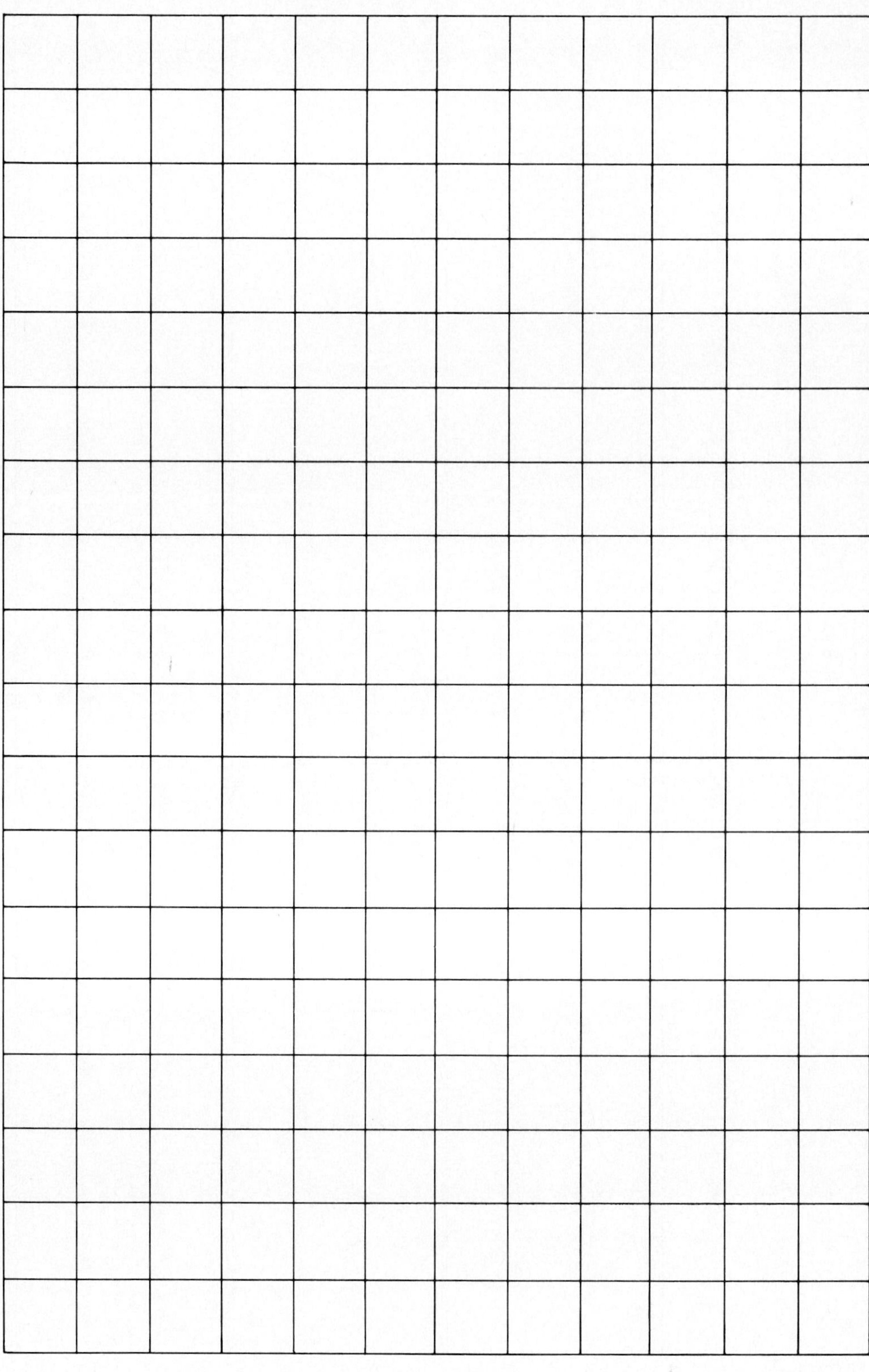

NAME

Introduction to Paper Pool

1. D C

A B

2.

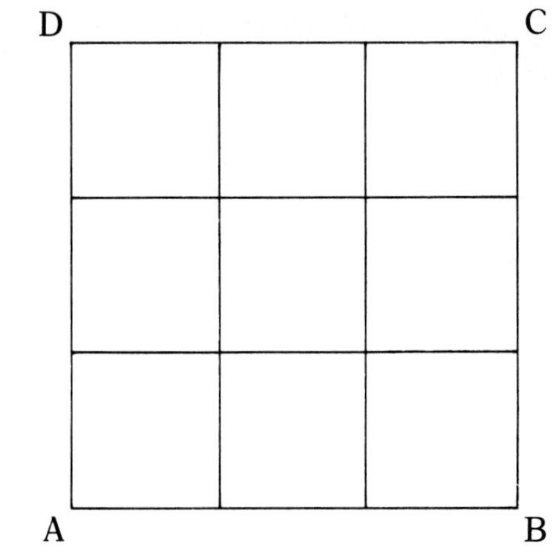

3. D C

A B

4.

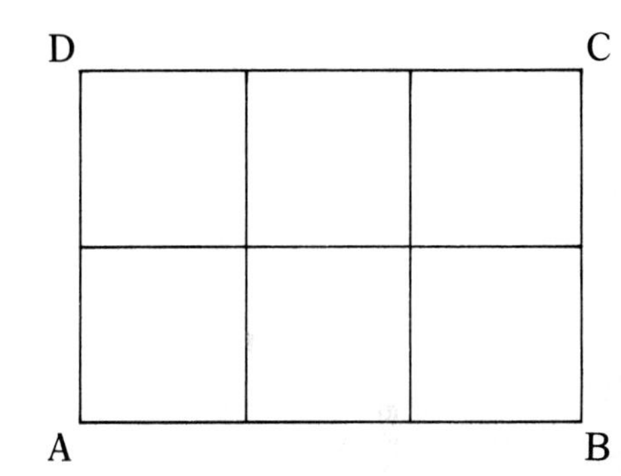

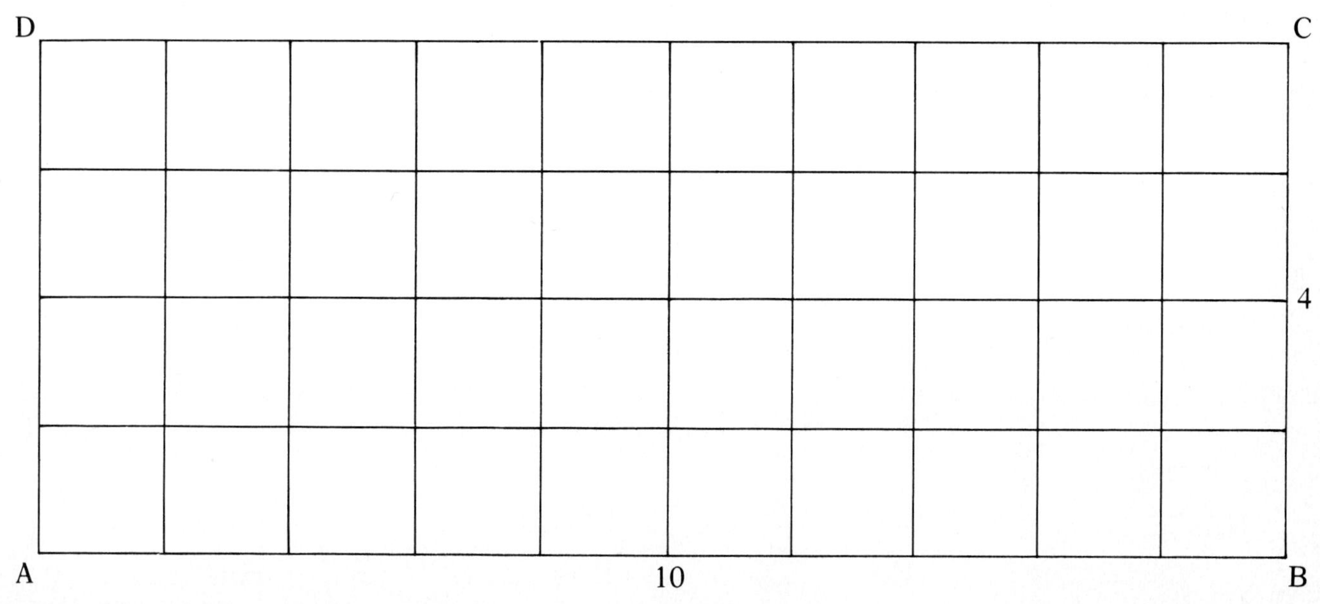

NAME ____________________

Paper Pool

Use a colored pencil to draw the path of the ball.

1. 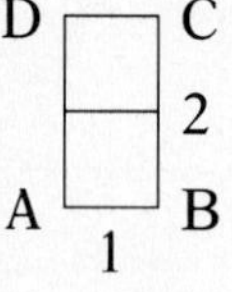

Corner ______

Hits ______

2. 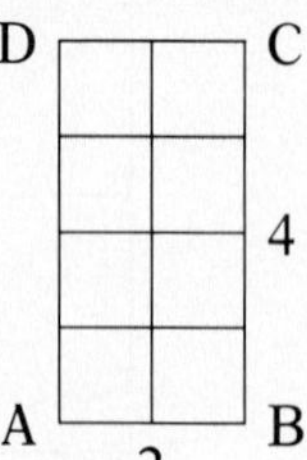

Corner ______

Hits ______

3. 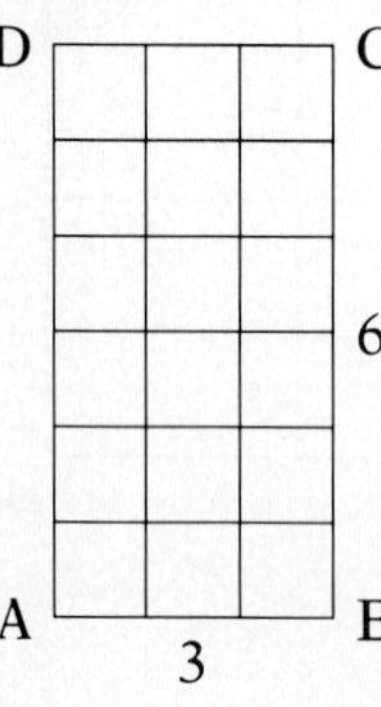

Corner ______

Hits ______

4. 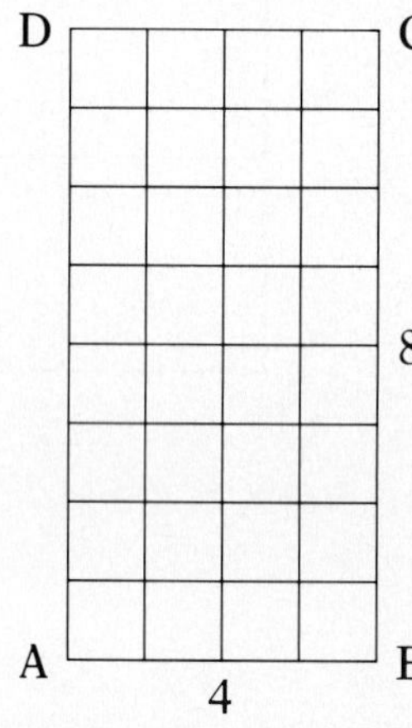

Corner ______

Hits ______

5. 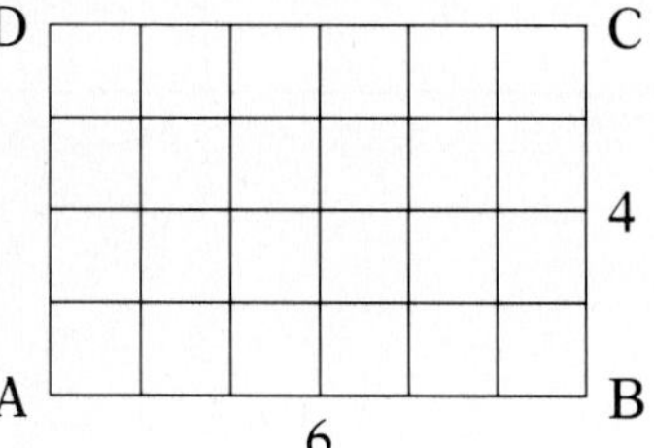

Corner ______

Hits ______

6. 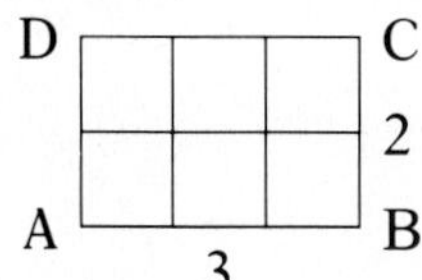

Corner ______

Hits ______

7. 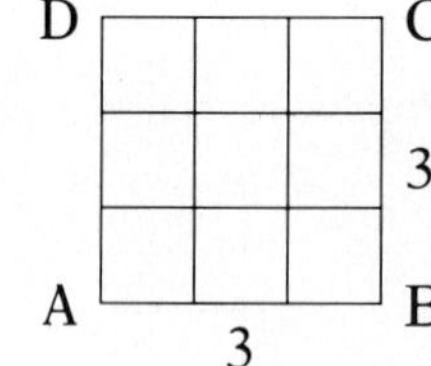

Corner ______

Hits ______

8. 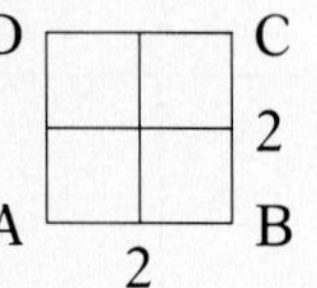

Corner ______

Hits ______

9. 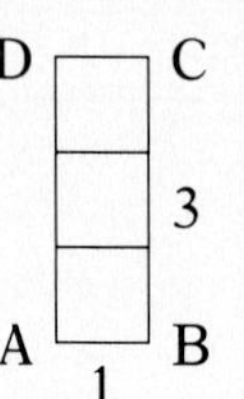

Corner ______

Hits ______

10. 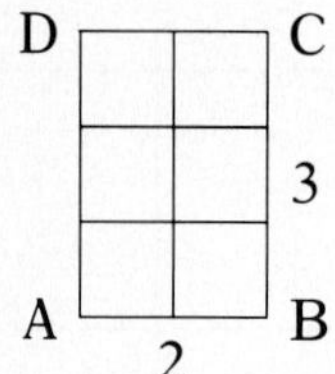

Corner ______

Hits ______

11. 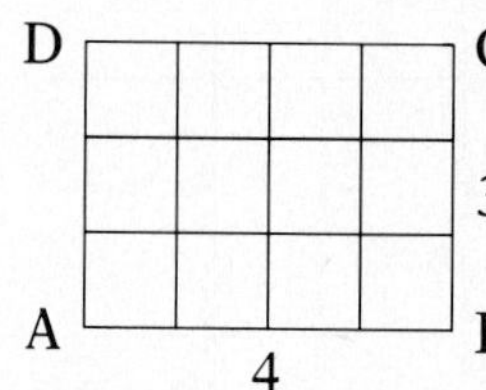

Corner ______

Hits ______

NAME ____________________

Paper Pool

Use a colored pencil to draw the path of the ball.

12.

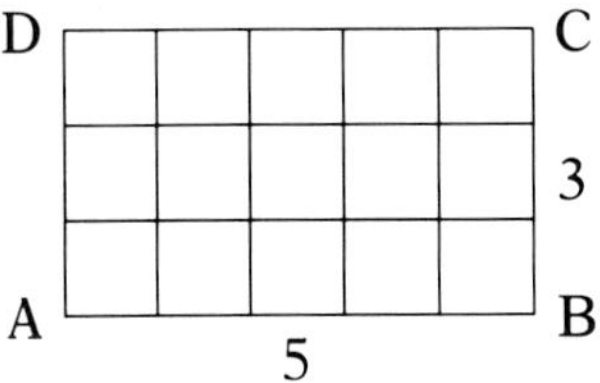

Corner ______

Hits ______

13.

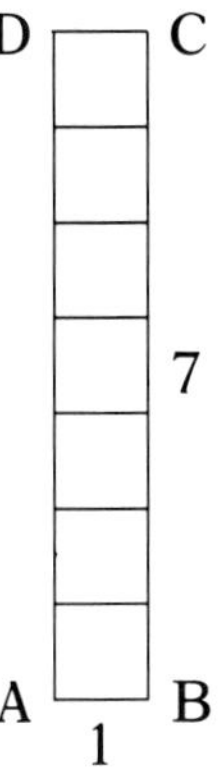

Corner ______

Hits ______

14.

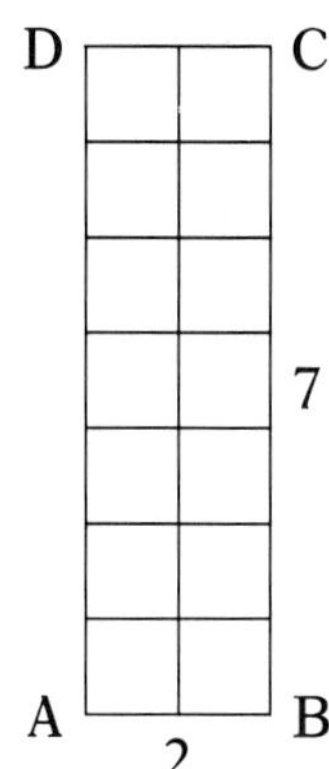

Corner ______

Hits ______

15.

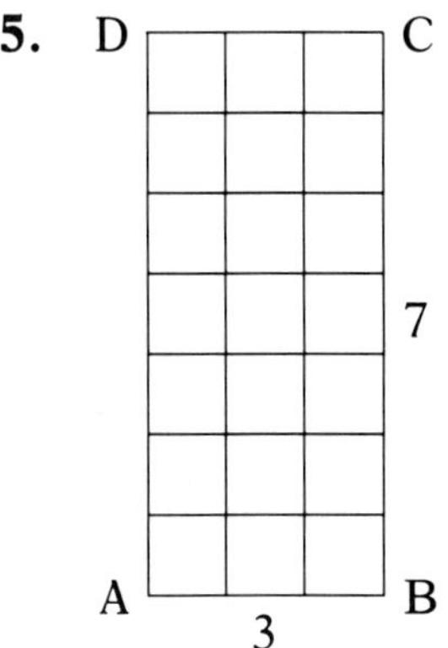

Corner ______

Hits ______

16.

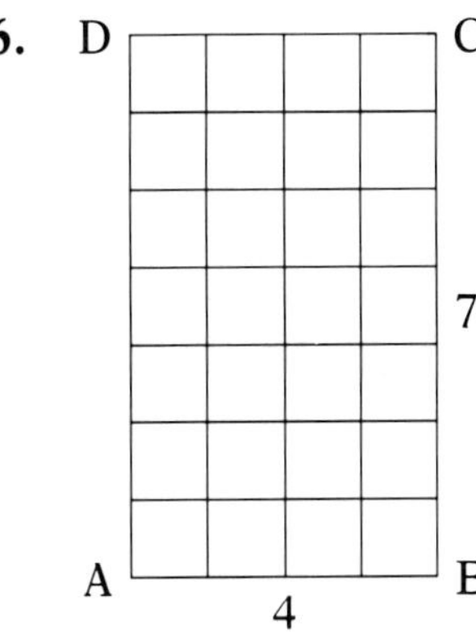

Corner ______

Hits ______

17.

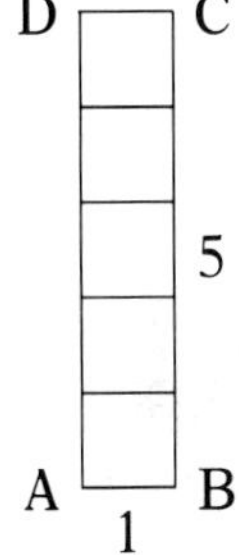

Corner ______

Hits ______

18.

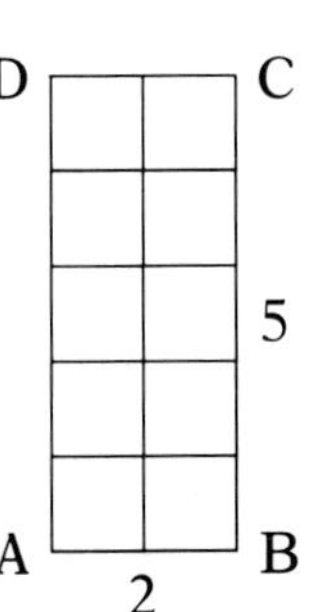

Corner ______

Hits ______

19.

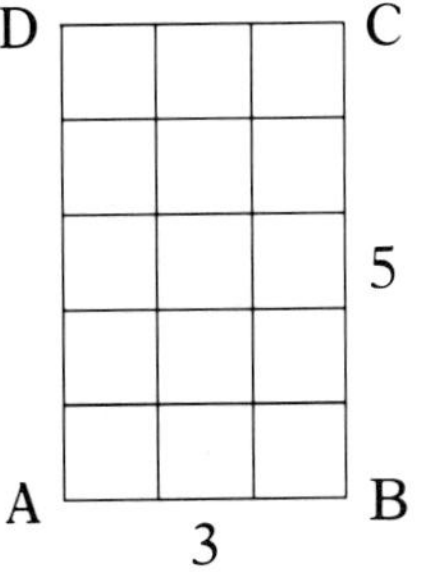

Corner ______

Hits ______

20.

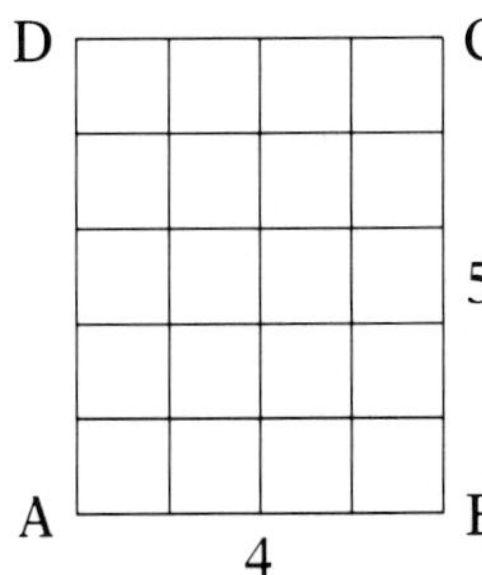

Corner ______

Hits ______

NAME ______________________

Paper Pool

Use a colored pencil to draw the path of the ball.

21.

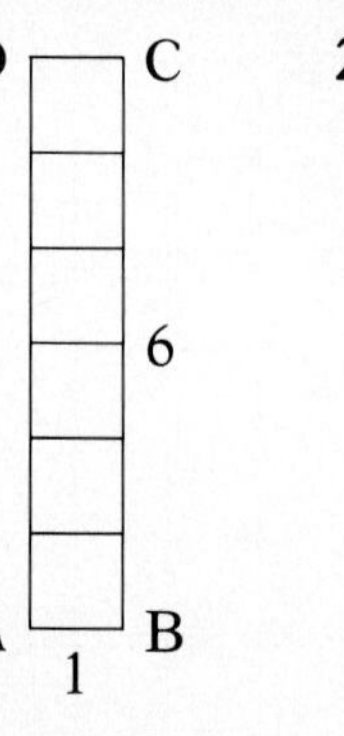

Corner ______

Hits ______

22.

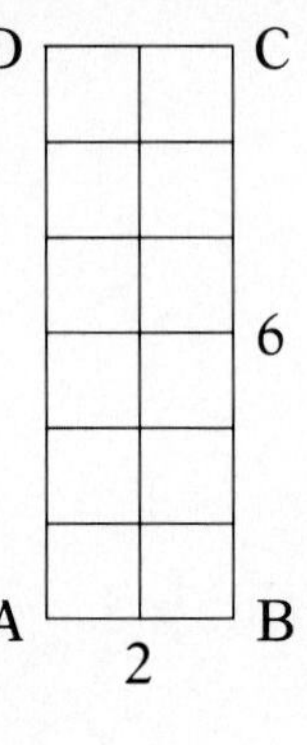

Corner ______

Hits ______

23.

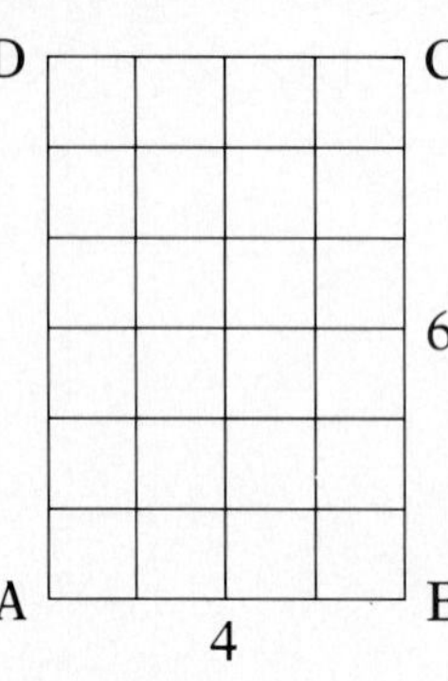

Corner ______

Hits ______

24.

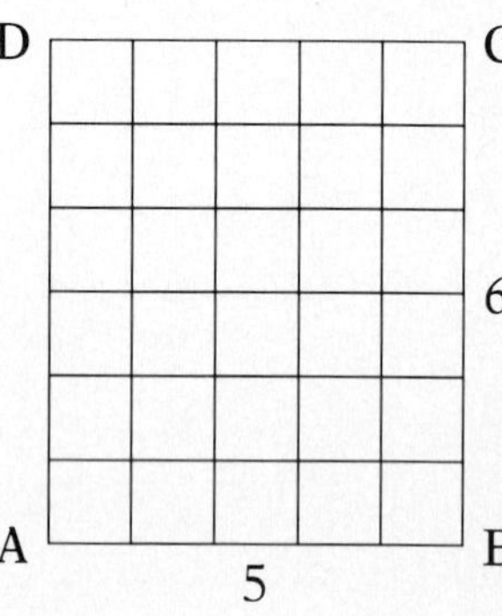

Corner ______

Hits ______

25.

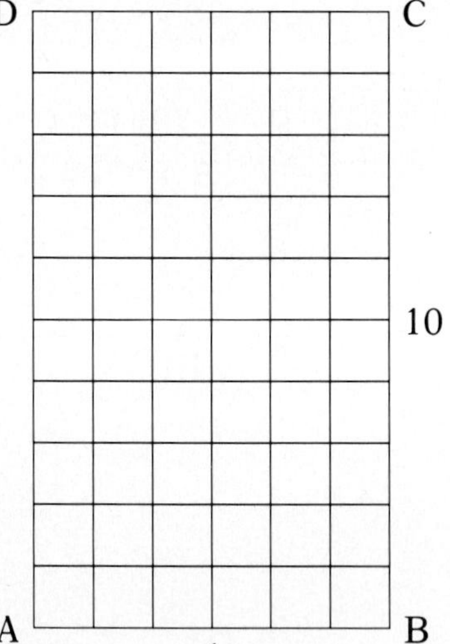

Corner ______

Hits ______

26.

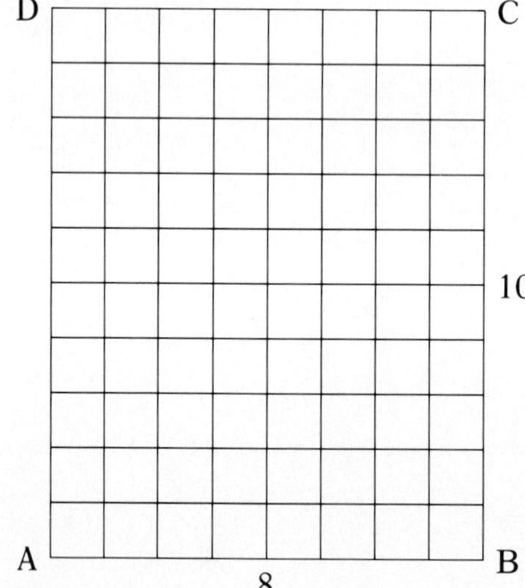

Corner ______

Hits ______

NAME ______________________

Record Sheet

Record the data from the pool tables in Worksheet 8-2 in this chart.
Remember to record *bottom edge* × *side edge.*

Corner where ball stops.	Total Number of Hits Including Start and Finish									
	2	3	4	5	6	7	8	9	10	11
B		4 × 2								
C										
D										

NAME

Supertables

Bottom Edge ______

Side Edge ______

Corner ______

Hits ______

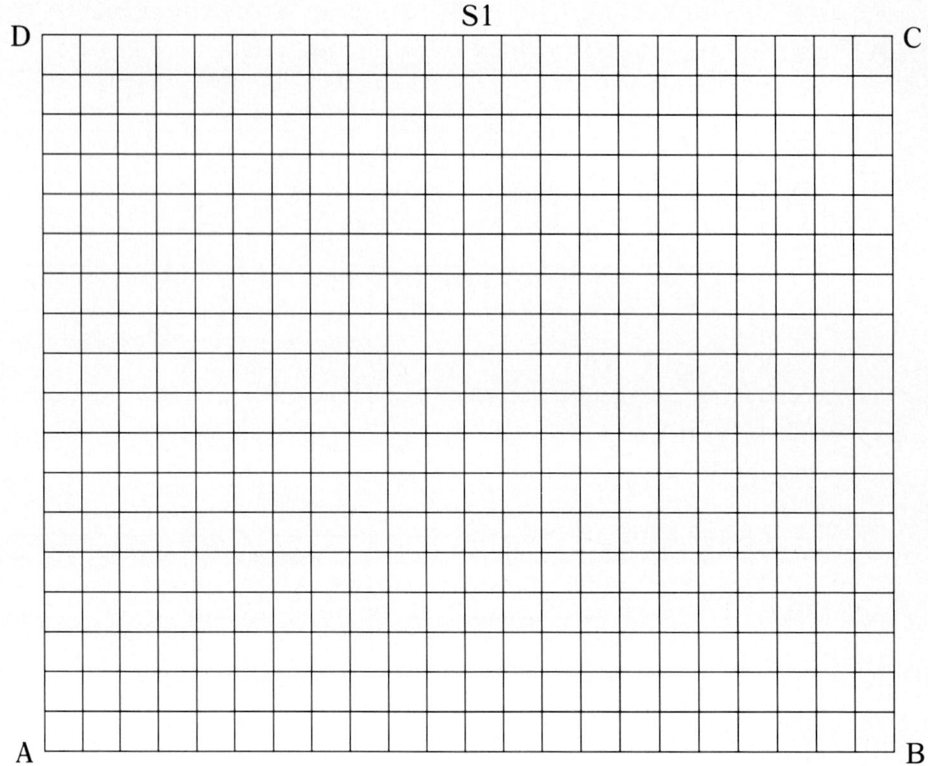

Bottom edge ______

Side edge ______

Corner ______

Hits ______

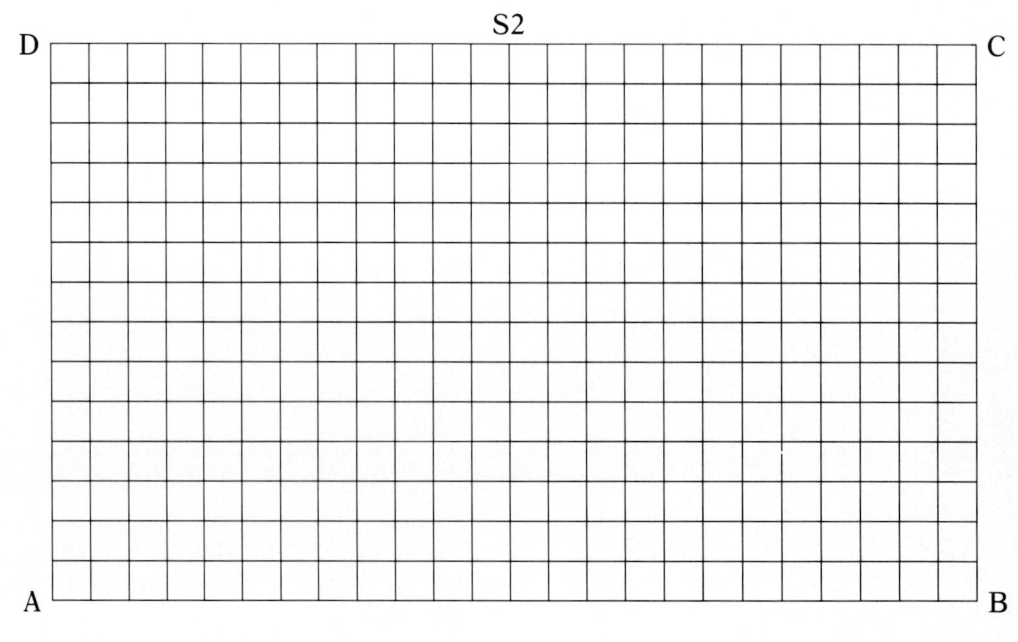

NAME ______________

Advanced Paper Pool

For each table, predict the final corner and the number of hits.

	Basic Table	Two Family Tables		Corner	Hits
1. 9 × 11					
2. 33 × 121					
3. 51 × 72					
4. 32 × 72					
5. 45 × 81					
6. 108 × 72					
7. 42 × 7					
8. 13 × 65					
9. 114 × 84					
10. 42 × 13					

11. Draw the path of the ball on this table. If the "6" is changed to "21", in order for the path to be the same, then the "8" should be changed to ______.

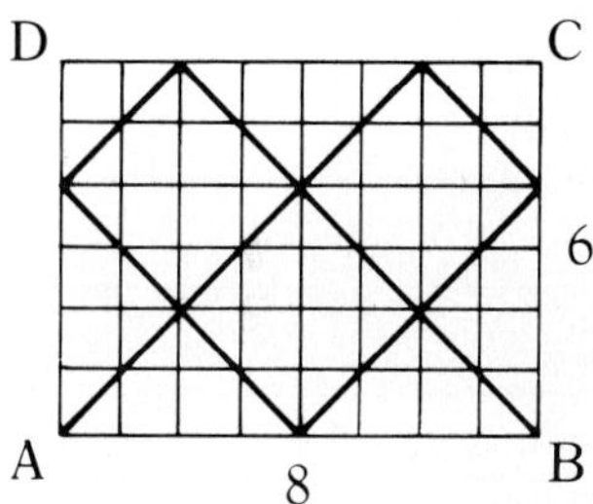

12. A table has bottom edge 14. If the side edge is 40, how many hits are there? ______

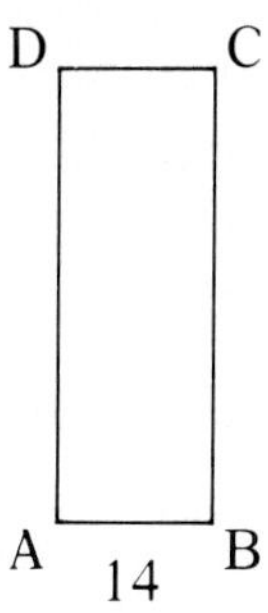

13. With the bottom edge 14, what is the smallest the side edge can be in order that there are at least 50 hits? ______

NAME ______________________

How Many Squares Are Crossed?

There is also a rule to predict how many squares the path crosses on a paper pool table.

For example: →⧄ This path crosses two squares. Use your Paper Pool worksheets and record the data from each table.

Size of Table	Number of Squares Crossed

Size of Table	Number of Squares Crossed

Mark the basic tables with a *.

What is the rule for the basic tables?

What is the rule for the other tables?

Does the rule for the other tables also work for basic tables?

How many squares would be crossed in a 24 × 36 table?

NAME ____________________

Review Problems

1. Find all the factor pairs for the following numbers.

a) 28 **b)** 72

2. What is the smallest number we have to check to be sure that we have all the factor pairs of the following numbers?

a) 60 **b)** 150

3. Decide whether the following numbers are deficient, perfect, or abundant.

a) 12 **b)** 28

c) 21 **d)** 64

4. Fill in the box with a number that will make each statement true.

a) $7 \times \square = 56$

b) $\square \times 3 = 36$

c) $14 \times 5 = \square$

5. Find the factors used to make the following Product Game.

4	■	6	9
10	14	15	■
21	25	35	49

6. Find all the divisors of the following numbers.

a) 30 **b)** 29

NAME ______________________

Review Problems

7. Mrs. Brown wrote a number on the chalkboard and said, "I know that to be sure I find all the factor pairs for this number I must check each number from 1–29." What numbers could Mrs. Brown have written? What is the smallest number Mrs. Brown could have written? What is the largest number Mrs. Brown could have written?

8. A number has exactly five factor pairs. If the number is less than 100 and larger than 50, what is the number?

9. Find the prime factorization for the following numbers.

a) 90 **b)** 71 **c)** 441 **d)** 246

10. a) What is the least common multiple of 6 and 15?

b) What is the greatest common factor of 6 and 15?

11. a) The least common multiple of a number and 12 is 60. What could the number be?

b) The greatest common factor of a number and 12 is 4. What could the number be?

12. a) In checking for all the primes less than 180, how far do you have to sift?

b) List all the primes between 200 and 220.

NAME

Review Problems

13. To find all the primes from 1 to N, you need only sift through 13.

    a) What is the smallest number N could be?

    b) What is the largest number N could be?

14. In paper pool, a certain table has dimensions 6×15.

    a) What pocket will the ball end up in?

    b) How many hits are there?

    c) How many squares are crossed?

15. If a pool table has a bottom edge of 12, give the dimensions of the side edge so that

    a) the ball ends up in pocket B.

    b) the ball has at least seven hits.

    c) the ball crosses at least 30 squares.

16. Draw all the rectangles with an area of 18.

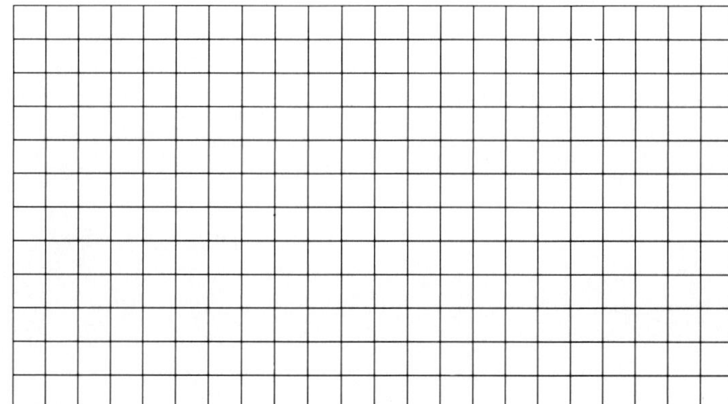

17. Every 4th day, pizzas are served on the school menus. Carrot salad is served every 6th day. If carrot salad and pizzas are served today, when is the next time they will be served together?

18. 150 baseball cards and 100 football cards are to be distributed evenly among a group of scouts. What are the possible numbers of scouts for this distribution to come out evenly?